Computing and Systems Technology Division 2018

Core Programming Area at the 2018 AIChE Spring Meeting and 14th Global Congress on Process Safety

Orlando, Florida, USA
22 – 25 April 2018

ISBN: 978-1-5108-6444-3

Printed from e-media with permission by:

Curran Associates, Inc.
57 Morehouse Lane
Red Hook, NY 12571

Some format issues inherent in the e-media version may also appear in this print version.

Printed by Curran Associates, Inc. (2018)

For permission requests, please contact AIChE
at the address below.

AIChE
120 Wall Street, FL 23
New York, NY 10005-4020

Phone: (800) 242-4363
Fax: (203) 775-5177

www.aiche.org

Additional copies of this publication are available from:

Curran Associates, Inc.
57 Morehouse Lane
Red Hook, NY 12571 USA
Phone: 845-758-0400
Fax: 845-758-2633
Email: curran@proceedings.com
Web: www.proceedings.com

TABLE OF CONTENTS

Some abstracts in this book refer to images that were not made available by the publisher

New Technology for Closed-Loop Multivariable System Identification, PID Control Loop Optimization and Advanced Process Control

J. Lepore and S. Howes

PiControl Solutions LLC, Houston, Texas, USA (e-mail: steve@picontrolsolutions.com).

Abstract: Ability to identify open-loop transfer functions using normal process data from a running plant is immensely useful. Open-loop transfer functions help to optimize PID tuning precisely, implement Advanced Process Control (APC) inside the plant's DCS or PLC, improve Model Predictive Control (MPC-DMC/RMPCT) performance by improving models and build dynamic process control simulator for training. This paper shows a new and novel method of using complete closed-loop data without any intrusive step tests for model identification. The technique identifies multivariable open-loop models with slave PIDs in auto or even cascade modes and then shows how to optimize PID tuning and improve MPC models. Software tools COLUMBO and PITOPS are used to illustrate various real plant examples.

Keywords: PID tuning software, PID tuning optimization, closed-loop multivariable system identification, dynamic modelling, Advanced Process Control (APC), Model Predictive Control (MPC), DMC (Dynamic Matrix Control)

1. INTRODUCTION

Most chemical plants today are controlled by a DCS (Distributed Control System) or a PLC (Programmable Logic Controller). Data historians store process data conveniently available in Excel format.

During normal plant operation, control room operators make setpoint changes required to adjust process conditions in order to achieve the desired product properties and production rates. PID controllers can be in auto or cascade modes with their setpoints adjusted by an operator, a DCS-resident APC scheme or a MPC like DMC or RMPCT.

Dynamic models are often expressed as transfer functions or step-response coefficients. Identification of open-loop dynamic models can be easily done by conducting intrusive step tests on a slave PID control loop in manual mode or even step tests on the setpoint in auto mode. Technology and tools for such open-loop tests and model identification has been used successfully for decades with good results.

However, simultaneous multivariable identification of open-loop dynamic models processing multiple inputs with slave PID control loops in cascade mode is far more difficult and often produces erroneous open-loop models.

This paper describes a new method using software called COLUMBO for identifying multivariable open-loop transfer functions using pure closed-loop data with slave PID control loops in cascade mode amidst high frequency noise, drifts, and unmeasured disturbances. No new intrusive open-loop tests are needed. The normal plant operating data is used.

2. OPEN-LOOP DYNAMIC MODELS

Almost 98% of all dynamic relationships in chemical processes can be characterized by zero-order, first-order, second-order, or open-loop unstable type transfer functions as shown in Figure 1.

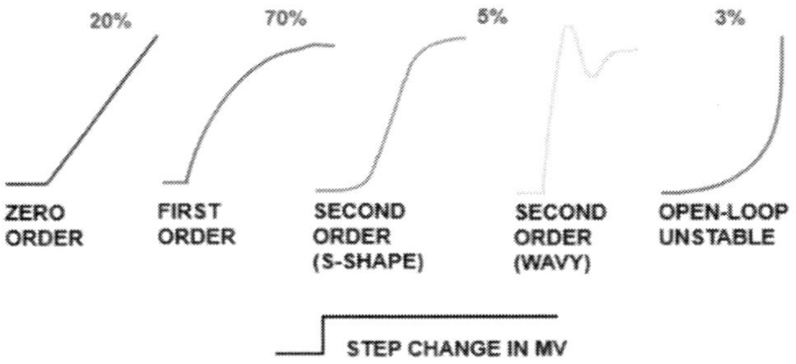

Figure 1. Open-loop dynamic models in chemical plants

The remaining 2% of models that show a very complex shape with inverse response and roller-coaster squiggly shapes are either not real, better off not used as a model in a MPC or could be represented as a reduced order model fitting one of the categories shown in Figure 1. Higher order models can be reduced to lower orders by extrapolating the dead time and minor shape modification without losing significant model prediction accuracy.

Current system identification algorithms and software work well for simple cases shown in Figure 2 below. However, in many cases such intrusive step tests are hard to conduct due to process sensitivity and interactions. Step tests may cause product properties to change unacceptably. Normal plant operation often involves ramping setpoints of slave PID control loops. With ramping there is no abrupt step test and conventional open-loop model identification methods may not be effective. During the normal plant operation, setpoints of multiple variables are often changed simultaneously. Multiple MVs can affect multiple CVs. Superimposed on these multivariable simultaneous changes are the menacing effects of unmeasured disturbances and noise. The new COLUMBO algorithm provides this new, novel breakthrough functionality illustrated in Figure 3.

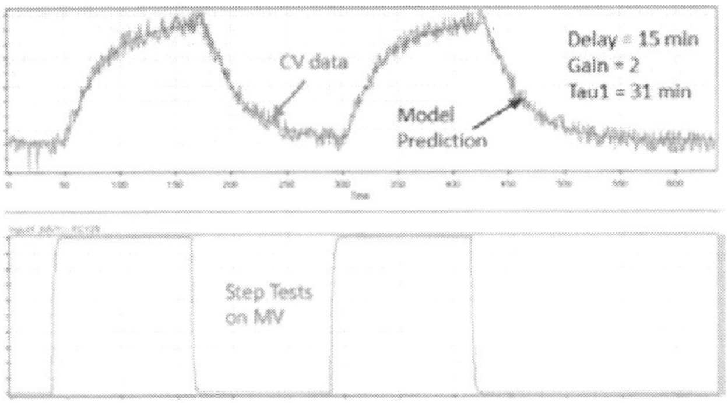

Figure 2. Conventional open-loop step tests

3. CLOSED-LOOP DATA FROM NORMAL OPERATION

Figure 3 shows control of gas composition in a reactor. CV data are gas composition. MV moves show slave PID flow setpoint changes in cascade mode made by the gas composition master controller in auto mode. This data is illustrative of typical and normal plant operation. At the very left is start-up followed by steady state and some planned setpoint changes. Strong unmeasured and unexplained disturbances are superimposed. With conventional system identification methods, this data would be considered "bad" or not "rich-enough" for identification of open-loop dynamic models. But the new COLUMBO algorithm can identify the open-loop transfer function as shown in Figure 3. The bottom-most trend shows the isolated unmeasured disturbance signal separated from the transfer function contribution.

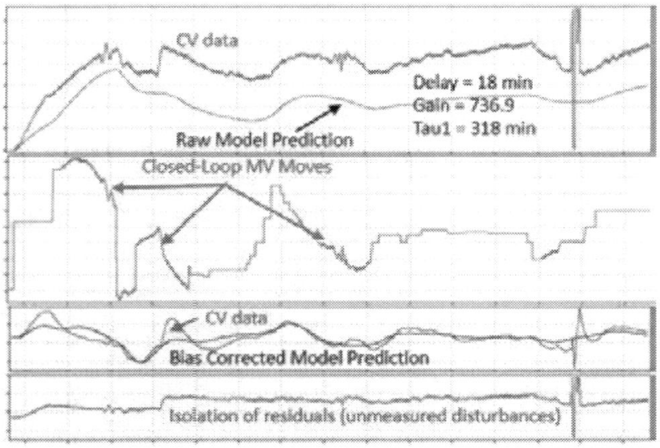

Figure 3. Closed-loop data from normal plant operation

Figure 4 shows another complex data set showing a master AC (analyzer controller) manipulating a slave FC controller. The changes in the CV value in the top trend is because of some setpoint ramping necessary for making the product grade changes. The data is oscillatory and full of unmeasured disturbances. The new COLUMBO algorithm can isolate the unmeasured disturbances (see the bottom-most trend in Figure 4) and identify the true open-loop transfer function model (delay = 15 min, process gain = 103.7 and first order time constant = 35 min). Conventional tools, algorithms and methodologies cannot make much out of such complex closed-loop data tainted by unmeasured disturbances and that's why COLUMBO is a true breakthrough in process control with its ability to identify the true open-loop transfer functions using such closed-loop data without any step tests.

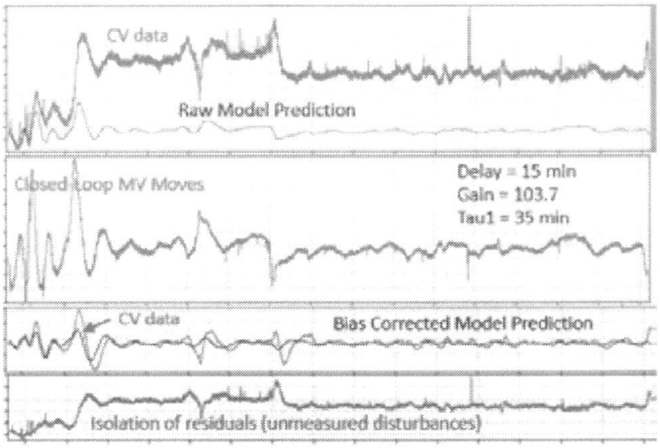

Figure 4. Closed-loop oscillatory data from normal plant operation

Figure 5 shows a slave PID in cascade mode. The slave PID's setpoint could be manipulated by a master PID, APC or MPC. The data shows pure normal plant operation. There have been no new step tests conducted for model identification. With the slave PID in cascade mode, COLUMBO algorithm can identify both open-loop transfer functions: one from control valve to slave CV and the second from slave setpoint to master CV. This ability is new, novel and distinguishes COLUMBO from other software tools.

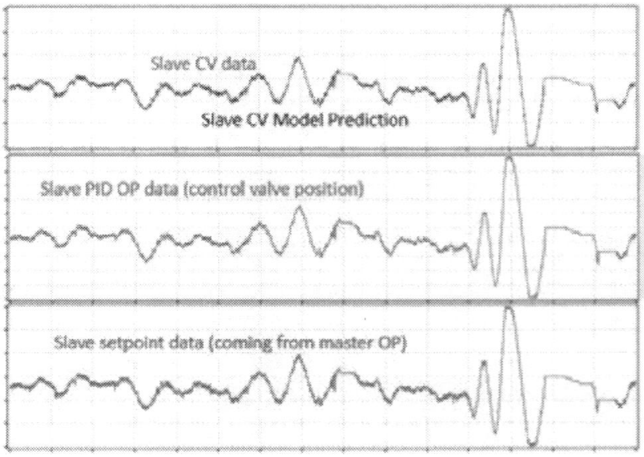

Figure 5. Master-Slave closed-loop data with master in auto and slave in cascade modes

Figures 6 and 7 show a spectacular example from a real plant distillation column. The column feed, reflux flow and reboiler duty change in a correlated manner – at the same time. A MPC (DMC) is on and makes changes to the slave setpoints with the slave PIDs in cascade/remote mode. Three open-loop transfer functions are simultaneously identified with completely closed-loop data. These identified open-loop transfer functions were used to modify models in the DMC. Control performance improved after the newly identified models were inserted into the DMC.

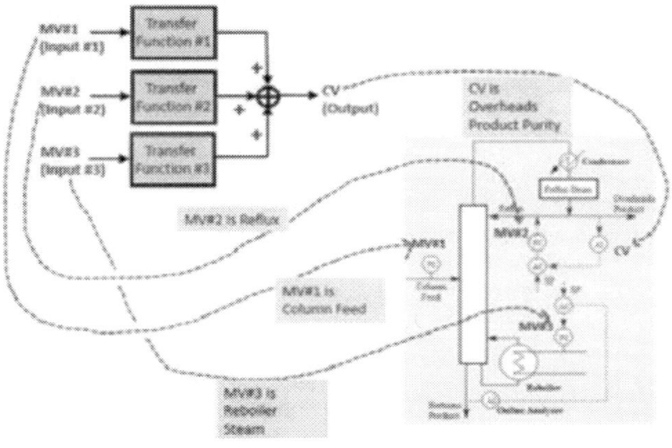

Figure 6. Multivariable closed-loop model identification

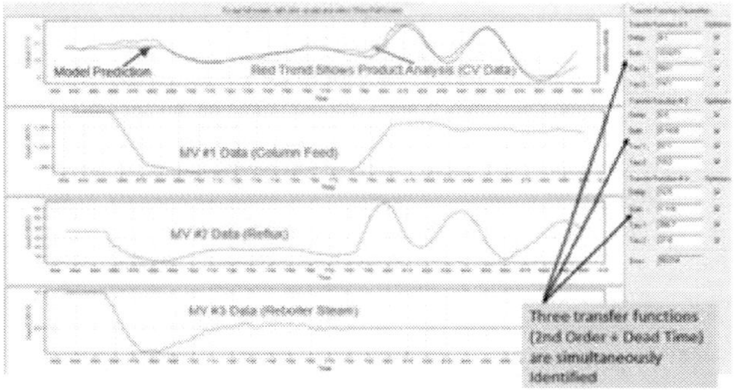

Figure 7. Multivariable model identification with complete closed-loop data

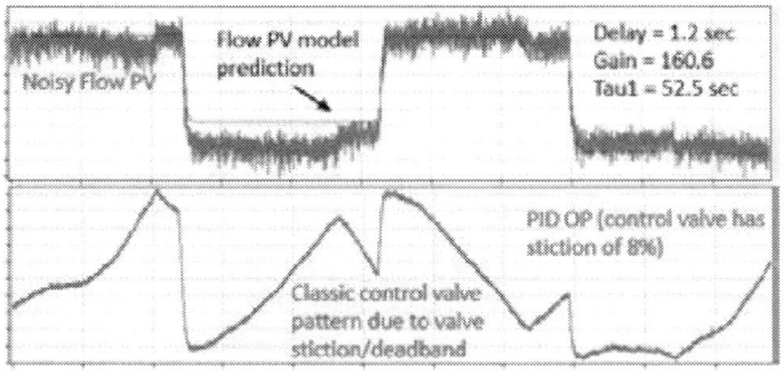

Figure 8. Identification of control valve stiction along with open-loop dynamic model for a flow PID control loop

Figure 8 shows another spectacular example showing the ability of the new COLUMBO algorithm to identify control valve stiction simultaneously in addition to the open-loop transfer function parameters using pure closed-loop data comprised of setpoint changes in auto mode. The ability of COLUMBO to identify control valve stiction with such normal operating data is extremely useful in chemical plants where old control valves can start deteriorating followed by loss of control quality. When slave PIDs deteriorate due to bad control valves, this also reduces control quality of an APC or MPC followed by lost profits and lost benefits.

4. ISOLATION OF RESIDUALS

Residuals are the unmeasured disturbances that creep into any process. A good example of an unmeasured disturbance is the change in heating value of fuel gas to a furnace for temperature control. Current system identification algorithms are sensitive to unmeasured disturbances and can generate false models. COLUMBO has the amazing capability to isolate the unmeasured disturbances and high frequency noise while determining the accurate dynamic models. Figure 9 shows a simulation example comprised of a single step test on the MV. The CV rises and settles and then due to an unmeasured disturbance, the CV creeps down a little. Using conventional model identification algorithms will produce a low process gain but COLUMBO produces the correct process gain of 5 in this example and isolates the unmeasured disturbance shown in the bottom trend. For ease of explaining and understanding, Figure 9 shows a simple example of an open-loop case based on simulation data, but this residual disturbance signal isolation is possible for multivariable cases also with pure closed-loop data.

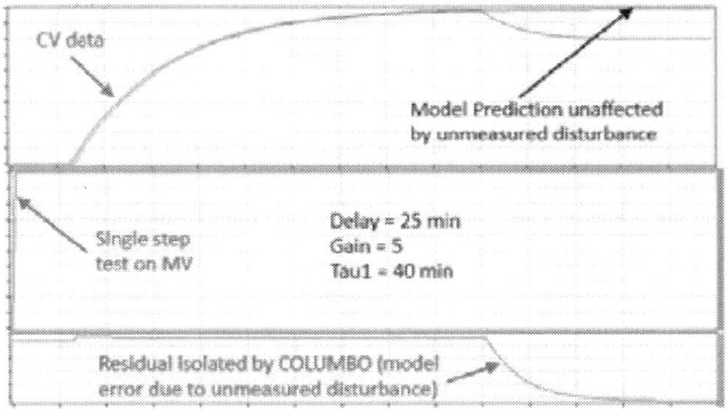

Figure 9. Isolation of unmeasured disturbances

5. PID TUNING OPTIMIZATION BASED ON CUSTOM SIMULATIONS

Most PID tuning done in the control room is still based on the trial-and-error method or using some heuristics like IMC (Internal Model Control), Ziegler Nichols Open-Loop/Closed-Loop, Cohen Coon or Lambda tuning. These methods were satisfactory for many years but in recent times, plants are built to be more efficient using many recycle streams, process interactions and complex designs making some of the old PID tuning methods hard to use effectively.

PITOPS tuning software offers new PID tuning algorithms and methodology that is easy to apply and effective on the newer, complex, and interactive plant designs. Instead of using generic heuristic equations for tuning, PITOPS minimizes the error between the setpoint and PV based on configured custom simulation. A custom simulation incorporates typical setpoint changes done by the operators or a cascade PID, APC or MPC. Typical disturbances and noise seen in the real DCS or PLC screens and trends can be easily configured in PITOPS.

Sometimes the goal for the PID is to respond well to step changes in the setpoint, sometimes may be the setpoint is ramped slowly. Sometimes setpoint of slaves need to respond fast based on their master PIDs. Sometimes, setpoints are never changed but the PID is often seeing severe disturbances that could be pulse, ramp, or sinusoidal waves. External or unmeasured disturbances that appear as sinusoidal waves could be due to interaction from neighboring PIDs. Conventional PID tuning also leans on relative gain analysis and building two loops interacting with each other. The new and novel approach from PITOPS allows you to configure a typical setpoint change (step/ramp or complex trajectory), typical disturbances and typical noise followed by minimization of the error between the setpoint and the PV of the PID. This approach works well for slaves, masters, multiple PIDs, constraint override PIDs, any type and any combination of a chain comprising of one or more PID loops. Figure 10 shows a PITOPS PID tuning optimization example where PITOPS minimizes the absolute error between the PV and setpoint for a simulation comprised of a setpoint change, ramp disturbance, pulse disturbance and noise in the sensor. The PITOPS optimizer allows imposing a rate-of-change limit on the PID's OP so that abrupt valve changes will not disturb downstream units. PID tuning parameters from this approach are markedly superior to trial-and-error and heuristic-type of PID tuning methods.

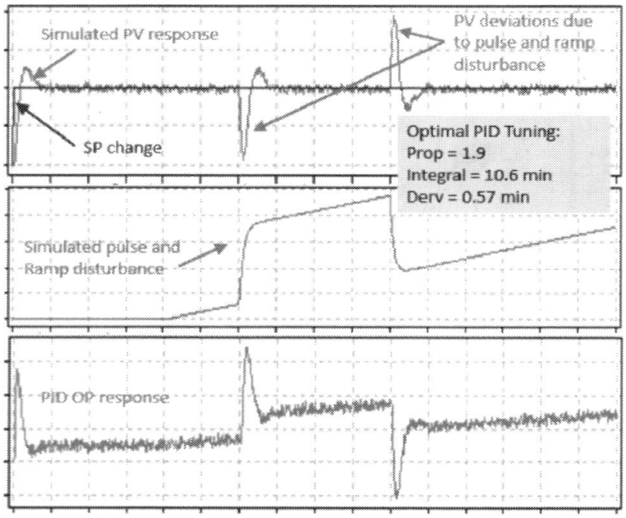

Figure 10. PID tuning optimization based on typical SP change, disturbances and noise as seen in real plant

6. ADVANCED PROCESS CONTROL (APC) INSIDE A DCS OR PLC

Knowledge of open-loop dynamic process models in the form of transfer function parameters can be used to design and implement powerful APC schemes inside a DCS or even a PLC. Many processes can benefit tremendously by implementing traditional APC using multiple cascade PIDs, selector-based override constraint control schemes, model-based control schemes, inferential control schemes and virtual sensors. These can be built in a DCS or PLC without the need for an MPC. Many engineers resort to selecting and implementing an MPC instead of a DCS/PLC-resident APC because they do not have practical tools for identifying open-loop transfer functions for various MV-CV pairs. With COLUMBO for identifying transfer functions and PITOPS for APC design, compact, robust, reliable, and easy-to-maintain APC schemes inside a DCS or PLC can be implemented at one fifth or less cost and effort compared to an MPC project. Many processes have a pseudo-diagonal control matrix where the density of the control matrix is low. Such processes are excellent candidates for a DCS or PLC-resident APC and can outperform an MPC and produce even more plant benefits and profits. See Figure 11.

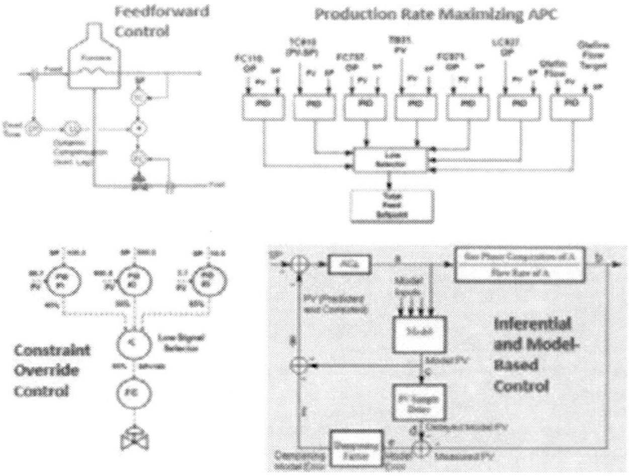

Figure 11. Traditional APC inside a DCS or PLC

7. IMPROVING MPC MODELS

MPCs like DMC, RMPCT, Connoisseur, PredictPro and others are based on dynamic process models based on step responses on various MVs. To generate good models, most MPC need uncorrelated step tests. Only one MV is stepped at a time. To identify the dynamic models accurately, holding each step anywhere from one-third to one-and-half times the time to steady state is recommended. Holding for such long-time periods after each move requires the step sizes to be small – typically only 1-3% of the prevailing values of the slave PID setpoints. Larger moves can upset the process.

But, when the MPC is on and working in closed-loop mode, the MV moves made by MPC can be significantly bigger compared to the 1-3% step tests during model identification. Nonlinearities and deviations from the simplified principle of linear superposition can cause the MPC models with larger and simultaneous MV moves to produce effects different from the identified open loop models. This is why many MPCs often need fine-tuning and model improvements. COLUMBO can convert any MPC model into a transfer function model and then use closed-loop data with an MPC running to improve the MPC models and the model prediction fit. This method is novel, unique and is the only method available for determining the real open-loop dynamic models based on large moves and correlated moves with several MVs moving simultaneously which is what happens when an MPC is on (active). See Figure 12 for an overview of COLUMBO.

8. COLUMBO AND PITOPS OPTIMIZATION SOFTWARE

Both COLUMBO and PITOPS are equipped with nonlinear constrained generalized reduced gradient (NC-GRG) optimizers. On top of the NC-GRG, PiControl has developed proprietary algorithms and code for working effectively with unmeasured disturbances, noise, closed-loop data with no step tests and processing of multivariable inputs. The technique has been successfully proven and tested with real plant.

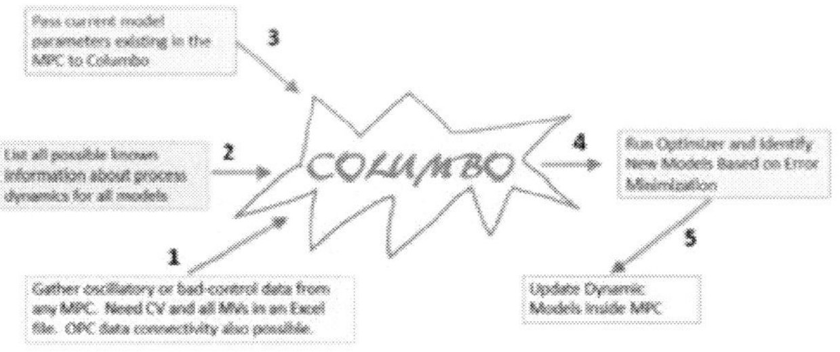

Figure 12. Steps using COLUMBO to improve MPC models

The COLUMBO approach offers a powerful unique capability that is truly a breakthrough. COLUMBO has the ability to allow the user to fix or set constraint limits on the various model parameters. In some cases, the time constants can be calculated based on chemical engineering knowledge and first principles. E.g., knowing the gas phase reactor volume and dividing by the flow rate gives the time constant. Or dead time could be determined based on some prior tests or based on pipe length, pipe diameter, gas flow and density.

COLUMBO allows fixing certain known parameters and then searches for all unknown parameters. This approach helps to identify more accurate dynamic model parameters. Often, error in process gain estimation in dynamic models is the root cause of MPC or APC problems and COLUMBO is an excellent new method for calculating accurate process gains.

Certain model parameters can also be fixed based on vendor data and information from experienced process operators and engineers. Only the COLUMBO algorithm allows reducing the uncertainty in the closed-loop optimization problem by incorporating process and dynamic knowledge based on various other sources and factors.

9. CONCLUSION

Most plants can generate Excel files containing MV, CV and FF (feedforward) data. The ability to identify open-loop transfer functions using complete closed-loop data with just the normal plant operation without any new intrusive step tests is a major improvement in the process control field.

System identification is inherently a complex area and COLUMBO makes the process easier and more successful compared to conventional competing approaches. COLUMBO offers the following powerful functionalities:

- Simultaneously identifies multivariable dynamic models
- Pure closed-loop data can be used from normal plant operation with some target changes without step tests
- Isolates unmeasured disturbances and noise
- Allows incorporating process knowledge, vessel geometry, chemical engineering first principles and vendor data and models into determining dynamic models and/or improving their accuracy.
- Works entirely in the time domain without need for complex math and no need for Laplace or Z (discrete) domain.

COLUMBO and PITOPS both use the NC-GRG (nonlinear constrained generalized reduced gradient optimization) method. Proprietary techniques developed by PiControl Solutions LLC eliminate the need for data pre-conditioning, preprocessing or normalization and other prerequisites required by existing known conventional methods like ARMAX and DMI (Dynamic Model Identification).

REFERENCES

Chen, C.-L. (1989). A Simple Method for On-line Identification and Controller Tuning. *AIChE Journal*, Volume 35, Issue 12, 2037-2039.

Ljung, L. (1999). *System Identification: Theory for the User*, 2nd ed., Prentice Hall, New Jersey.

Escobar, M., Trierweiler, J.O. (2013). Multivariable PID Controller Design for Chemical Processes by Frequency Response Approximation. *Chem. Eng. Sci.*, Volume 88, 1–15.

Bucz, Š., Kozakova, A., Vesely, V. (2015). Ultimate Data Based Robust PID Design for Performance. 8th IFAC Symposium on Robust Control Design, *IFAC-PapersOnLine*, Volume 48, Issue 14, 221-228.

Wang, L., Freeman, C., Rogers, E. (2017). Experimental Evaluation of Automatic Tuning of PID Controllers for an Electro-Mechanical System. 20th IFAC World Congress, *IFAC-PapersOnLine*, Volume 50, Issue 1, 3063-3068.

Gao, X., Shang, C., Huang, D., Yang, F. (2017). A novel approach to monitoring and maintenance of industrial PID controllers. *Control Eng. Pract.*, Volume 64, 111-126.

Pitops user manual, "Process Identification & Controller Tuning Optimizer Simulator", version 1.4.5.0.

Sharmaa, R., Glemmestadb, B. (2013). On Generalized Reduced Gradient method with multi-start and self-optimizing control structure for gas lift allocation optimization. *J. Process Contr.*, Volume 23, Issue 8, 1129–1140.

Peng, H., Ozaki, T., Toyoda, Y., Shioya, H., Nakano, K., Haggan-Ozaki, V., Mori, M. (2004). RBF-ARX Model-Based Nonlinear System Modelling and Predictive Control with Application to a NOx Decomposition Process. *Control Eng. Pract.*, Volume 12, Issue 4, 191–203.

513410 A Stylized Trend Analysis Approach for Process Monitoring and Fault Diagnosis

Abstract:

A stylized trend analysis was developed to identify recurring or standard operating procedures (SOP) such as equipment switching or maintenance from plant process history data to detect faults and near misses due to operator errors or equipment failures. Trend based fault detection and diagnosis systems have not been implemented widely in the chemical process industries because known fault scenarios are typically required. Known fault scenarios were not required for the stylized trend analysis because faults and near misses were identified by comparing trends with normal SOP transition responses. Large volumes of historical data were processed automatically allowing corrective action to be taken prior to an incident. The stylized trend analysis was demonstrated to detect unsteady-state transition faults on historical distributed control system (DCS) data simulated for a continuous process ethylene plant dryer switching operation.

513357 Development of Process Operating Policies and Control Using Flowsheet Simulators, Plant SCADA, and CFD

Abstract:

Computational fluid dynamics models and flowsheet simulations have been developed for process control and operation of the Valrico Advanced Wastewater Treatment Plant in Valrico, FL. Using plant SCADA data, oxidation ditch models were created with Ansys Fluent to analyze nitrification-denitrification reaction kinetics and solid residence times--as well as abstract information on dead-zones and regions of mass transfer limitations--in order to generate and tune theoretical reactor models in BioWin software for flowsheet simulation. Plant operation was further expounded by simulating influent and operating conditions (e.g., recylce and aeration rates) for the entire plant to accurately model current effluent behavior and compositions. Lastly, historical influent SCADA data was used to evaluate model performance and develop operating policies for irregular flow periods and non-traditional plant operation and maintenance.

513148 Refinery Front-End Crude Scheduling with Consideration of Unit Maintenance

Abstract:

The front-end crude scheduling (FECS) is very important for the entire supply chain management of refineries. It is used to tackle the crude movement problem covering crude vessel unloading at ports, temporarily storing at storage tanks, multi-oil transferring through long-distance pipelines, crude blending in charging tanks, and continuously feeding to refineries under specifications. Through FECS, an optimal crude movement schedule considering both profitability and operability can be identified. In previous studies, however, such a schedule was normally obtained under the scenario that all process units were under normal and deterministic operating conditions without the consideration of any possible abnormal situations such as the demand of certain unit maintenance. Note that the crude supply process is a highly unit-intensive and interactive process coupling both batch and continuous operations under heavy duties. Thus, the unit depreciation and performance degradation cannot be neglected, and maintenance should be regularly planned and conducted [1,2]. For instance, all equipment needs to be periodically checked, repaired and even replaced; otherwise, equipment failures may results in significant economic loss, and severe safety and environment problems. Some research papers have been published on various uncertainties during normal operations, such as shipping delays, demand change and tank malfunctions [3,4,5], these studies focused on the identification of post remedial measures for some uncertainty scenarios, instead of proactively considering unit maintenance operations in advance. Therefore, there still lack of FECS studies with the consideration of the planning of unit maintenance. In this paper, a novel FECS model combined with the planning of unit maintenance has been developed. It can provide optimal schedules for both front-end crude movement and maintenance operations (starting time and duration) toward the minimal operational cost. The efficacy and performance of the proposed model has been tested and demonstrated through various case studies.

References

[1] Amaran, S., Sahinidis, N., Sharda, B., Morrison, M., Bury, S., Miller, S., & Wassick, J., (2015) Long-term turnaround planning for integrated chemical sites. Computer and Chemical Engineering, 72, 145-158.

[2] Vieira, M., Varela, T., Povoa, A., (2017) Production and maintenance planning optimization in biopharmaceutical processed under performance decay using a continuous-time formulation: A multi-objective approach. Computer and Chemical Engineering, 107, 111-139.

[3] Xu, J., Qu, H., Wang, S., & Xu, Q., (2017) A new proactive scheduling methodology for front-end crude oil and refinery operations under uncertainty of shipping delay. Industrial & Engineering Chemistry Research, 56, 8041-8053.

[4] Li, J., Misener, R., & Floudas, C., (2012) Scheduling of crude oil operations uncer demand uncertainty: A robust optimization framework coupled with global optimization. AIChE Journal, 58, 2373-2396.

[5] Zhang, S., Wang, S., & Xu, Q., (2015) A new reactive scheduling approach for short-term crude oil operations under tank malfunction. Industrial & Engineering Chemistry Research, 54, 12438-12454.

513269 Process Stream Routing and Multi-Plant Operation of Industrial Wastewater Treatment Facilities

Abstract:

The authors have developed operating strategies for energy cost minimization and effluent regulation of an interconnected system of advanced wastewater treatment plants in Hillsborough County, FL. Inner-loop optimization has been performed on individual plant-wide transient operation of energy-intensive processes, while outer-loop optimization has been designed for influent and process stream manipulation and transmission between sites via plant-to-plant interconnects, transfer pumps, and lift stations. Forecasting models have been generated using artificial neural networks for inter-plant operability, and rules have been generated for supervisory decision making.

513168 Dynamic and Manufacturing Scheduling Approaching the Maximum Profitability Applications of Air Separation Units

Abstract:

Dynamic and Manufacturing Scheduling Approaching the Maximum Profitability Applications of Air Separation Units

Song Wang, Honglin Qu, and Qiang Xu

Dan F. Smith Department of Chemical Engineering

Lamar University, Beaumont, TX 77710, USA

Abstract

Cryogenic air separation processes use different in boiling points of the components to separate air into the desired products oxygen, nitrogen and argon. Unlike other chemical processes, the cost of feedstock (air) of an ASU is trivial; meanwhile, the energy supply of the entire process is totally in the charge of electricity power. ASUs are also characterized by fluctuating operating conditions to respond to changing product demands [1]. Although it is not much necessary to improve current ASU designs since there are quite proven techniques of cryogenic air separation design existing in industries; however, there are still significant potentials for improving their profitability in terms of optimal scheduling and operation ASU manufacturing.

Currently, many air separation plants are operated in a dynamic economic environment [2]. The electricity price could frequently fluctuate according to real time marketing (RTM) demands in different regions. In this paper, a general methodology based on dynamic simulation and scheduling of an ASU has been developed, which can optimize the ASU operation with respect to the electricity cost to achieve the best profit as well as highly efficiency. It contains three merits: (i) considering the entire ASUs dynamic simulation case study with detailed control strategy; (ii) optimal scheduling for maximize process profitability under uncertainties; (iii) smart production decision based on marketing supply-demands relationship. First, the methodology develops reliable dynamic simulation models under various production modes (capacities) and studies the dynamic transition performance among those production modes. Second, the optimal scheduling and transition will be conducted to maximize the plant profitability under the electricity cost pattern. Finally, the optimized ASU schedule and operations will be fine turned and validated by rigorous dynamic simulations. The developed methodology can be further extended to help ASUs for the predictive and proactive smart manufacturing.

Keywords: Air Separation Unit, Production Scheduling, Maximum Profitability, Smart Manufacturing

References

[1] Rui, H; Victor M. Zavala; Lorenz T. Biegler *. Advanced Step Nonlinear Model Predictive Control for Air Separation Units. Journal of Process Control. www.elsevier.com/locate/jprocont.

[2] Li, T*; Thierry, R; and Marc, B. Production Scheduling of Air Separation Processes. Air Liquide, Newark, DE 19702; Air Liquide, Champigny-sur-Marne, France; Air Liquide, Madrid, Spain.

513553 Hybrid Simulation of a Dynamic System

Abstract:

Process simulators are the foundation of modern process synthesis, plant design, and production analytics. These versatile tools enable engineers to effortlessly perform mass, energy, and species balance calculations for complex networks of unit operations, under both steady-state and dynamic conditions.

A challenge arises when a process has special features that are not readily accommodated by the simulation platform. Examples include unit operations where internal gradients influence the connection between inputs and outputs, the use of species and/or operating ranges for which thermodynamic relations are not available, and vendor proprietary equipment such as membranes. For these cases, process simulators provide interface capabilities to enable users to incorporate their own custom models into their process calculations, thereby greatly extending their applicability.

In this talk, we explain and demonstrate a hybrid simulation for a dynamic process involving real-time information exchange between the CHEMCAD® process simulation platform and MATLAB®, a standard tool for mathematical modeling and technical computing.

OPTIMISING EDC CRACKER OPERATION TO ENHANCE VCM PLANT ECONOMICS

Sreekumar Maroor[1], Stepan Spatenka[1], Wipada Lertpukpon[2], Pheeraya Kityanyong[2], Kritsada Chotiwiriyakun[2], Nattawat Tiensai[2] and Renato Wong[1], (1)Process Systems Enterprise Ltd., London, United Kingdom, (2)SCG Chemicals Co., Ltd., Bangkok, Thailand

Contact: Sreekumar Maroor, Tel: +44 20 8563 0888,

Email: s.maroor@psenterprise.com

1. Introduction

Ethylene dichloride (EDC) is converted to vinyl chloride monomer (VCM), the key building block for polyvinyl chloride (PVC), by energy-intensive thermal cracking. EDC cracking is carried out in furnaces which typically operate at a conversion of 55-60% and have a run length of more than a year. EDC cracking also produces small amounts of impurities such as butadiene, methyl chloride, ethyl chloride etc., whose concentrations need to be maintained below specified values to avoid operational issues in the plant.

Coke is another by-product of EDC cracking and its deposition on cracker tube walls leads to increased pressure drop and high tube skin temperatures, with the consequence that eventually the furnace needs to be taken offline for decoking.

The paper describes a digital operations approach that involves the development and application of rigorous models for the EDC cracking furnace for monitoring and optimising operation. The furnace model is based on detailed radical-based cracking kinetic mechanism and includes kinetics for coke deposition in the coils. Such a model, once validated against plant data, serves as a valuable tool to predict key performance indicators (KPIs) such as conversion, yields and coking rates. It can also be used for optimisation of the cracker operation to improve the VCM plant economics.

The approach has successfully been used in a project with SCG Chemicals, Thailand to model and optimise their EDC crackers. The cracking and coking kinetic models were tuned to plant data and the validated model is then used to perform optimisation of the EDC cracker operation. The optimisation results identify potential for significant savings due to operation at higher conversion, without adversely affecting key impurity yields and run length.

2. EDC cracking at SCG Chemicals

A schematic of the EDC cracking unit utilized at SCG Chemicals is illustrated in Figure 1. The EDC cracking unit consists of three zones: 1) preheat zone, 2) convection zone, and 3) radiant zone. In the preheat zone, liquid EDC from oxychlorination process and/or import is fed at the top of unit and preheated by hot flue gas. The preheated EDC is then evaporated by a vaporizer. In the convection zone, the vapour EDC is further heated by hot flue gas making EDC to be ready for pyrolysis reactions. Finally, the heated EDC is fed to the radiant zone where pyrolysis reactions take place. The VCM product (480-520 °C) is sent to a transfer line exchanger (TLE) unit for terminating the reactions (i.e., preventing the production of coke and undesired by-products at high temperature).

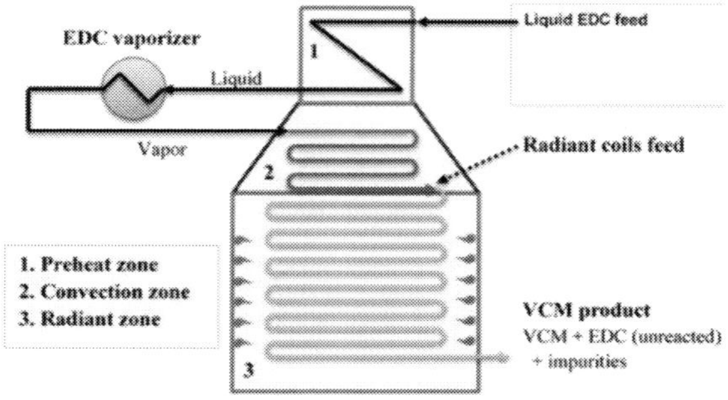

Figure 1 : Schematic of EDC Cracking Unit utilized at SCG Chemicals

2.1 Challenges to operational excellence

Key Performance Indicators (KPI) of EDC cracking unit commonly are conversion and run length. For the same VCM production, operation at higher conversion reduces the recycle flow in the plant and results in lower consumption of energy. However it can result in lower run lengths and higher levels of impurity composition in the plant. Higher levels of impurities can cause increased loads on the distillation columns and lead to significant inefficiencies. At SCG Chemicals, an operational target is to maximise the cracking conversion under a controlled level of impurities, while maintaining the unit's run length. In order to accomplish all dimensions of this target, a calculation tool is required for simultaneously monitoring, analysing, and predicting a set of optimised operating conditions. Such a tool would enhance operational competitiveness of the company in faster and cheaper manner.

3. Advanced process modelling to enable operational excellence

EDC conversion, yield of VCM and impurities, and coking rates are key factors for profitable operation of VCM plant. However, these are not normally measured directly because of the challenging physical environment. Hence the plant operators cannot see the effect of operational changes on these key variables. A model which can accurately predict these key variables valuable insight to the plant operators to take informed operational decisions. Such a model can also be used for identifying optimal operating policies to improve the VCM plant economics.

3.1 EDC cracker modelling in gPROMS ProcessBuilder

SCG's EDC cracker is modelled in gPROMS ProcessBuilder using PSE's cracking furnace models. Figure 2 shows the details of the model.

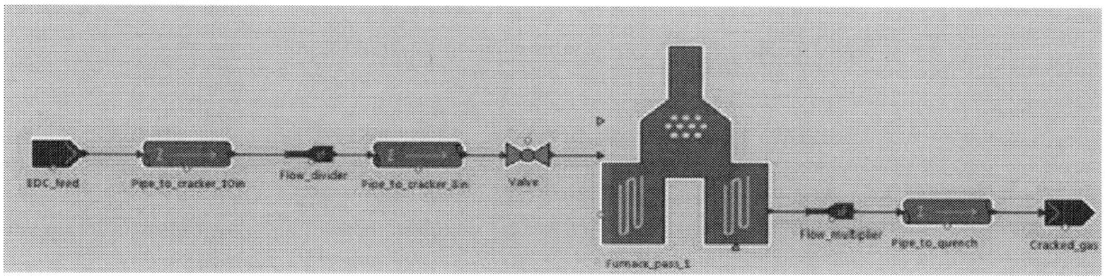

Figure 2 : Model of SCG's EDC cracker system

The model is configurable and allows detailed representation of the furnace geometry including radiant coils, adiabatic sections and transfer line exchangers (TLE) as shown in Figure 3

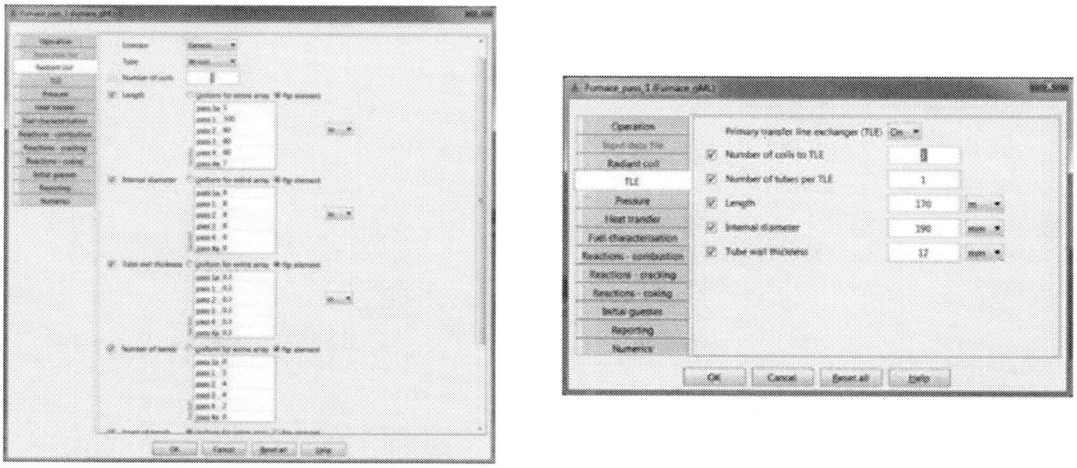

Figure 3 : Configuration of radiant coil geometry and TLE within cracking furnace model

Typically, process gas temperatures are measured at a few points along the length of the radiant coil. This information is provided as input and used by the model to approximate the temperature profile along the length of the radiant coil. This approach allows accurate representation of temperature profile in the radiant coil without relying on rigorous modelling of the fire box. Specification of gas temperature profile and other heat transfer related inputs within the furnace model is shown in Figure 4.

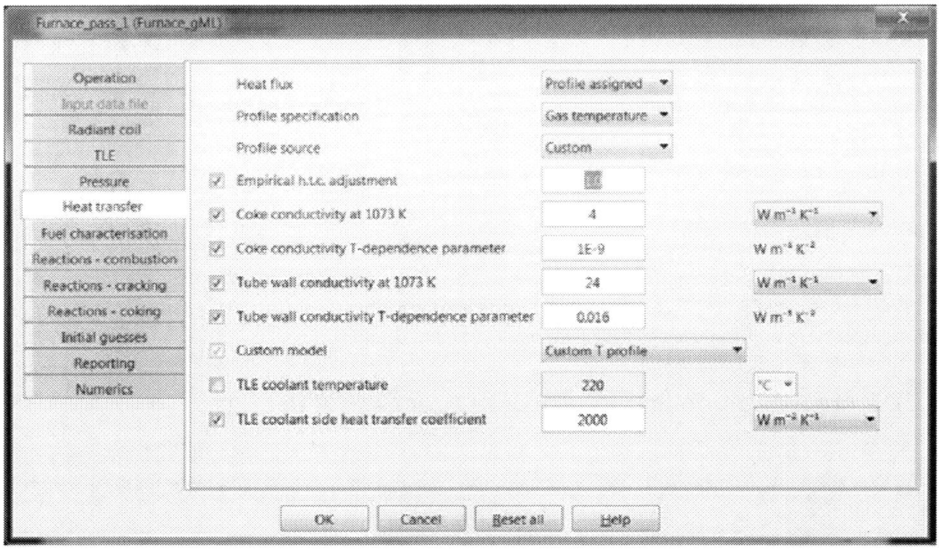

Figure 4 : Specification of heat transfer inputs within furnace model

Model calculations for conversion and yields are based on detailed radical based cracking reaction kinetics [1] involving 47 species and 108 reactions covering most components of industrial relevance. The model framework allows extension of the reaction scheme and tuning of the kinetic parameters based on available data using state of the art parameter estimation capabilities of gPROMS. Specification of cracking kinetics within the furnace model is shown in Figure 5. Coking rates are calculated using a coking kinetic model formulated based on information available in literature [2] [3].

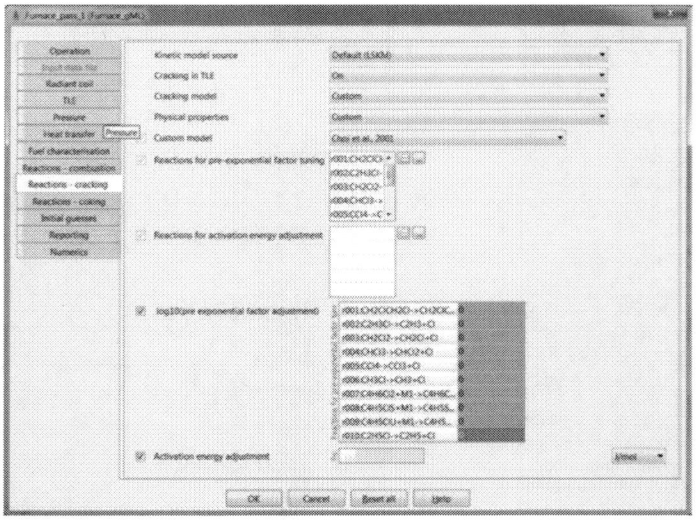
Figure 5: Specification of cracking kinetics within furnace model

3.2 Cracking kinetic validation and tuning

At SCG, cracked gas compositions at cracker outlets are not measured. Hence cracking kinetic model validation is performed by comparing the model predictions to cracked gas composition data established on the basis of overall plant mass balances. Comparison of model prediction and plant data for EDC and VCM compositions at the cracker outlet showed a good match and demonstrated the capability of the model to predict EDC conversion and VCM yield accurately. The prediction of by-product compositions showed scope for improvement by kinetic tuning. However, the plant data available for kinetic tuning was limited and as such it is not suitable for estimating all kinetic parameters in a large kinetic scheme. It was necessary to ensure that the information embedded in the original kinetic scheme is preserved as much as possible during the kinetic tuning process. Hence parameter estimations were set up such that changes to the kinetic parameters during model tuning are minimal from their prior values reported in literature.

Figure 6 show the comparison of model prediction and plant data for the composition of EDC and VCM at the cracker outlet showing the capability of the model to predict EDC conversion and VCM yield accurately. EDC and VCM compositions are predicted within 0.5 % (relative).

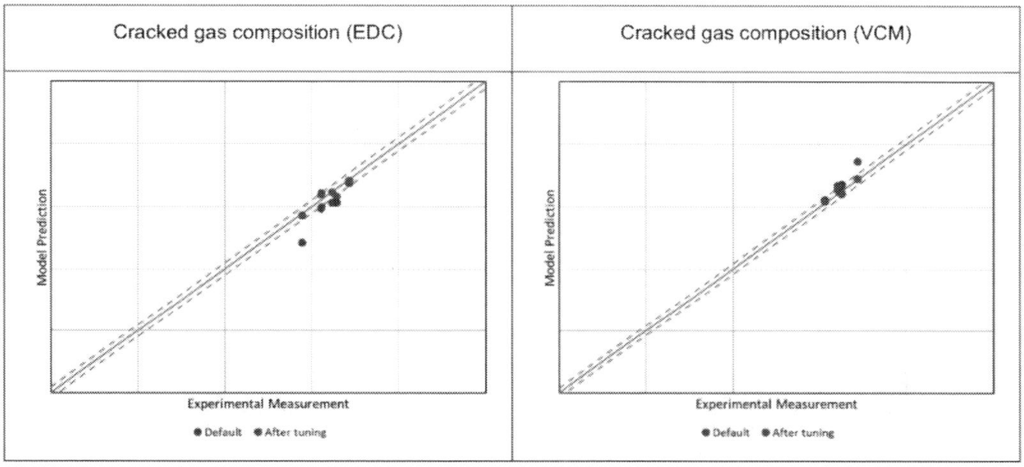

Figure 6 : Cracked gas composition prediction of EDC and VCM - comparison to plant data

Error bar: ±0.5% (relative)

Figure 7 and Figure 8 show the comparison of model prediction and plant data for the composition of impurities. Kinetic tuning helped to improve the prediction of key impurities.

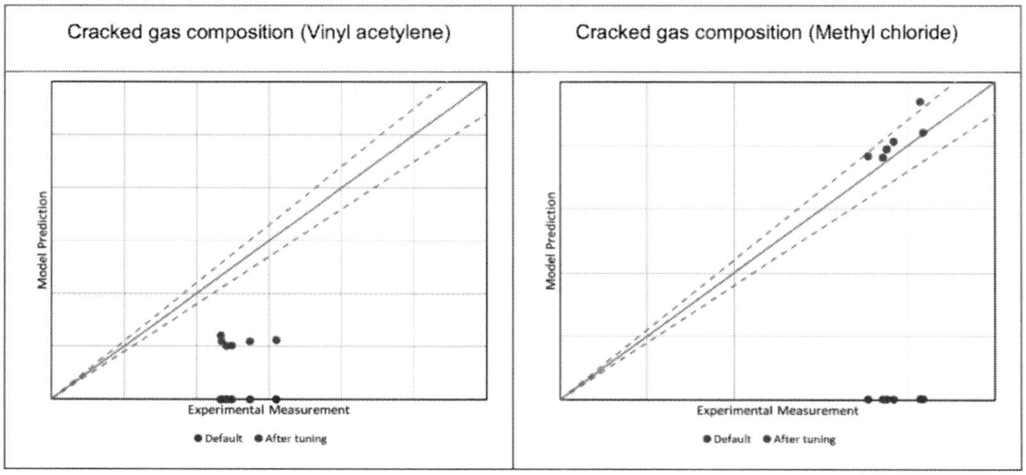

Figure 7 : Cracked gas composition prediction for key impurities – Comparison to plant data

Error bar: ±10% (relative)

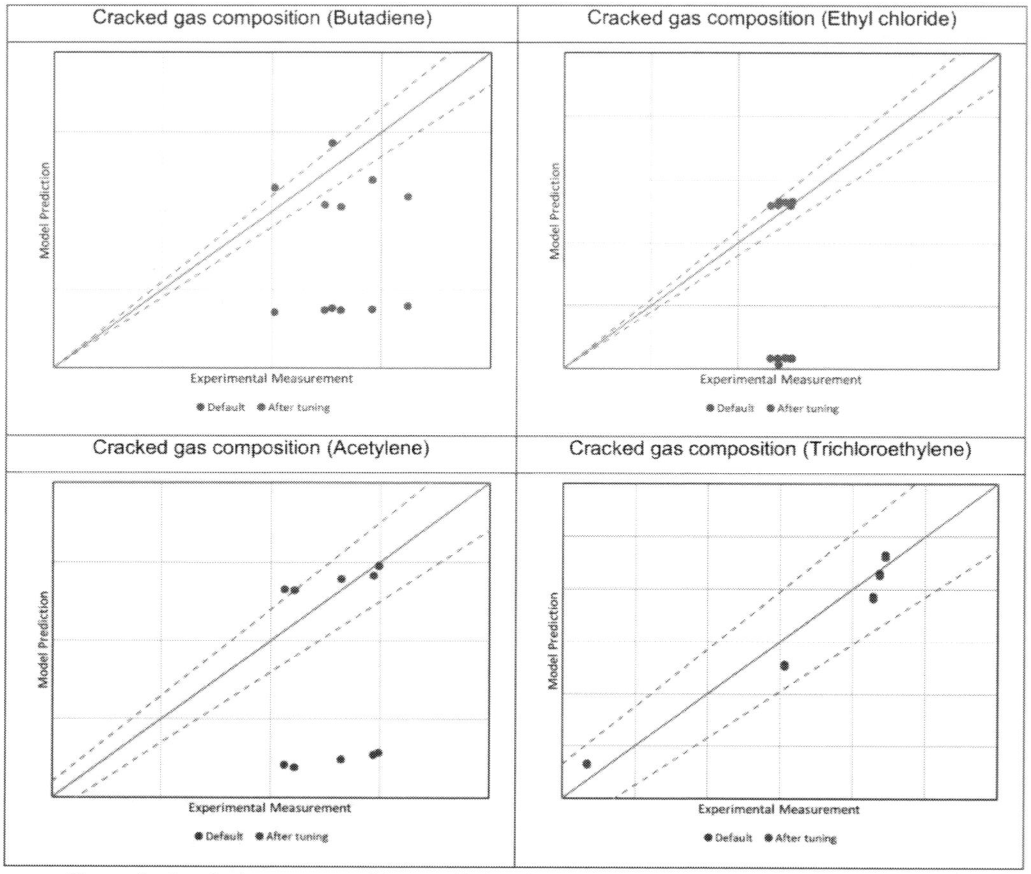

Figure 8 : Cracked gas composition prediction for key impurities – Comparison to plant data

Error bar: ±10% (relative)

3.3 Coking kinetic tuning

Coke is a by-product of EDC cracking and it deposits on the wall of the radiant coil and TLE, resulting in lower heat transfer coefficient and higher pressure drops. Optimisation of cracker operation requires a predictive coking kinetic model. The parameters of the coking kinetic model were estimated using plant data from two cycles. Key measurements used for coking model tuning include pressure drop across the furnace and coke thickness at the coil outlet measured at end of run. Tube metal temperature data was found to be unreliable and hence not used for coking kinetic parameter estimations. Figure 9 shows the comparison of pressure drop and VCM flow rate predictions to plant data for the two cycles considered for model tuning.

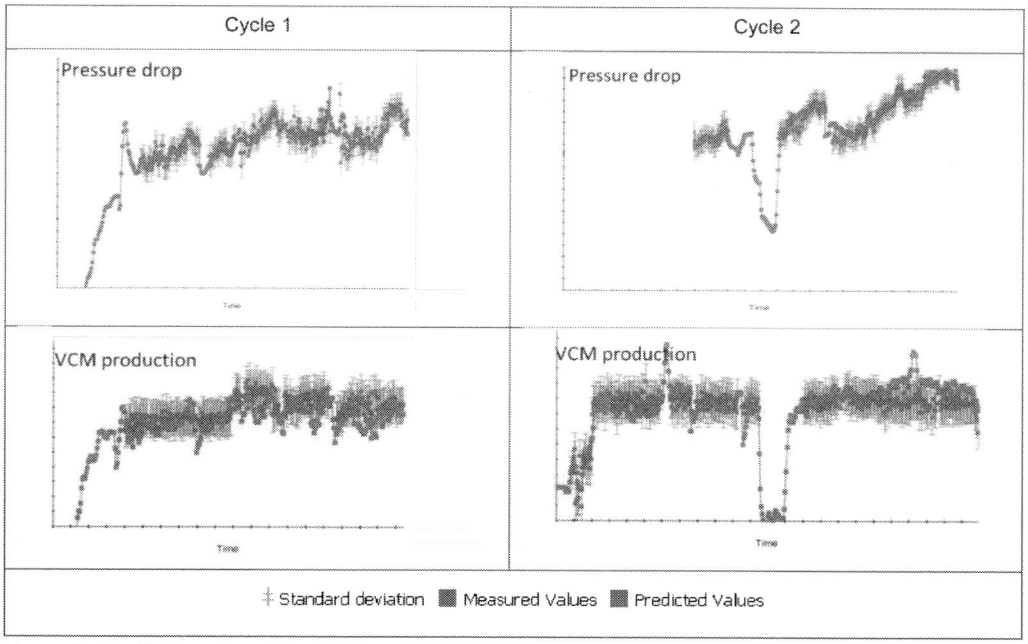

Figure 9 : Prediction of pressure drop and VCM production over the cycle and comparison to plant data

3.4 Optimisation

After validating the cracking and coking kinetics to available plant data, an optimisation problem was formulated to maximise the profit subject to plant constraints. More details of the objective function and key constraints are summarised in Table 1 (see Nomenclature for symbol definitions).

Objective function
Maximise $Profit = \frac{365}{Rl+SD}\left(\int_0^{Rl}\left[R_{VCM}(t) - C_{EDC}(t) - C_{fg}(t) - C_{Steam}(t)\right]dt - C_{dc}\right)$
Decision variables
• Operating temperatures in radiant coil • Cracking promotor concentration in feed
Key constraints
• Desired VCM production • Pressure drop at end of run • Maximum impurity compositions in cracker effluent

Table 1 : Optimisation set up

In this phase of the project, the model for optimisation did not include rigorous models of distillation columns and other units in the plant. Steam consumption quantities used in the objective function are calculated by relating it to the flow rate entering the respective unit using past data.

3.4.1 Optimisation results

SCG has defined a base case for the optimisation study. Figure 10 shows the results from an initial optimisation run (blue curves) and compares it to the base case (orange curves). In both cases VCM production is held constant. The optimiser suggests an alternate temperature profile, which allows to operate at higher conversion while maintaining the desired VCM production. The optimal temperature profile lowers the coking rate and hence improve the run length by ~ 25%. It improves the selectivity and hence results in lowering the unit cost of EDC. Optimised case also offers savings on fuel gas and steam consumption. Overall this brings in significant improvement to the VCM plant economics.

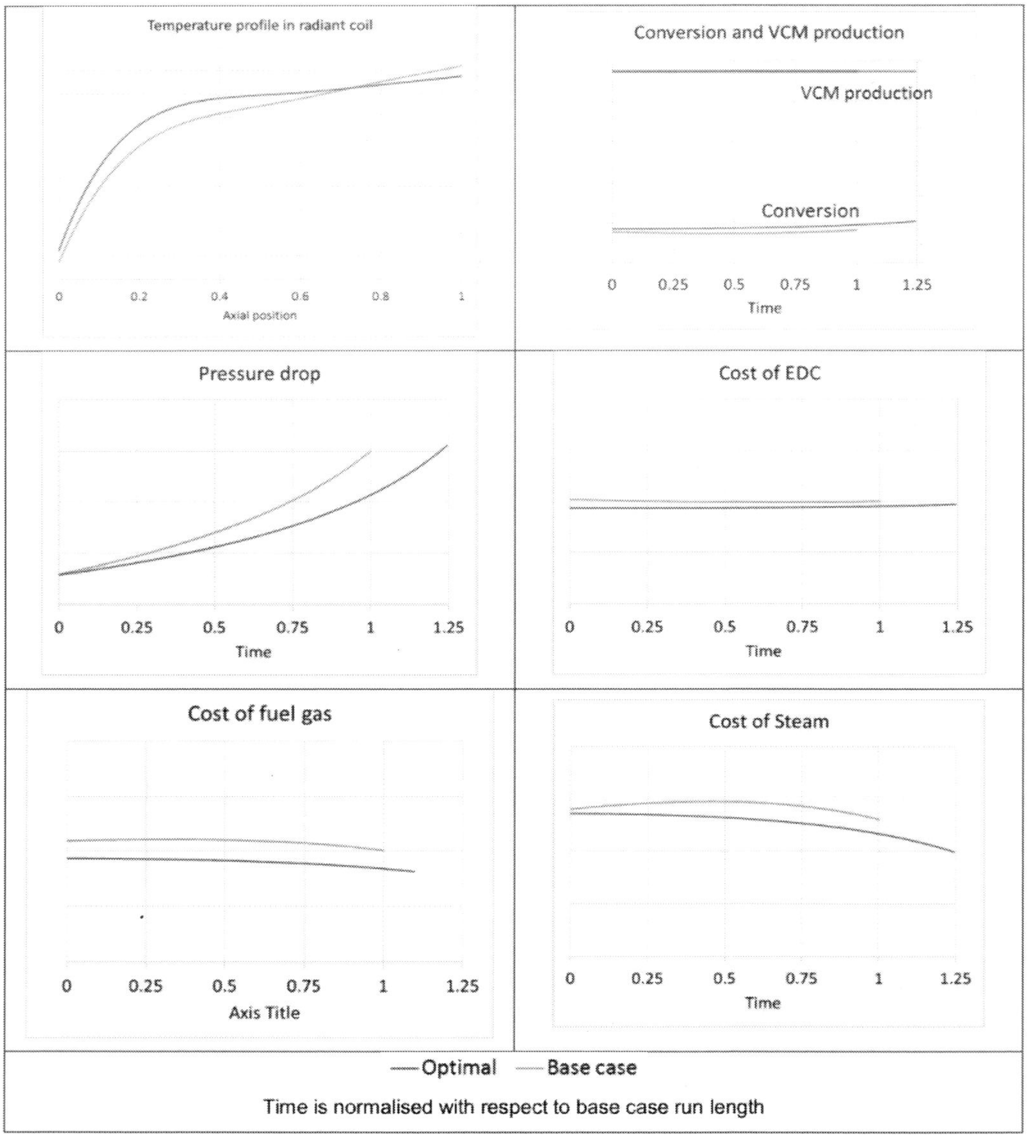

Figure 10 : Optimisation results

4. Application of model at SCG to drive operational excellence and benefits

The developed model and optimisation was successfully applied at one of the EDC cracking units in the company. The model was used to prescribe improved operating conditions (e.g., process temperature profile along the reactor coil), and suggest guidelines for daily operation, to help achieve operation at higher conversion without violating the plant constraints. SCG performed a trial implementation of the optimisation results in the plant. In this first implementation, without any plant equipment modification, the cracking conversion was increased by ~ 0.5%, resulting in an estimated cost saving of ~1%. Increasing conversion to higher levels requires plant equipment modifications. In future, with the necessary plant modifications, SCG targets an increase in conversion of 3%, which would result in more significant cost savings.

In addition, the model was instrumental in assessing the effect of operating conditions on KPIs and in preventing the risk of abnormal level of main impurities resulting from inappropriate operating conditions. Figure 11 shows the concentration profile of key impurities along the length of the radiant coil and TLE. Such plots provide detailed insight about the region within the coil where the rate of impurity production is highest. Manipulating process gas temperatures in these regions helps to manage the impurity production within allowable limits.

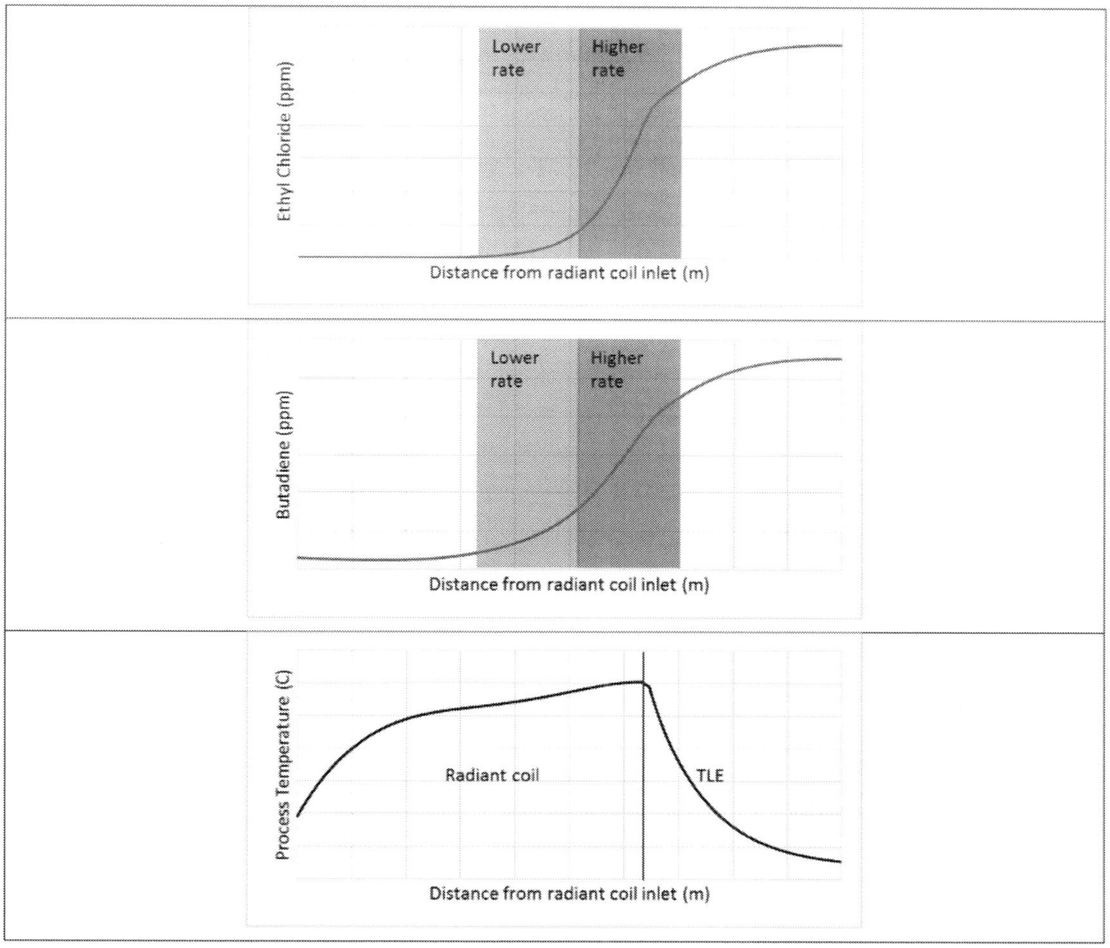

Figure 11 : Application of model in studying effect of process temperature to key impurities

5. Conclusions

A rigorous first-principles model of an EDC cracking furnace can be used effectively to optimise furnace performance and give new insights into operation. Such a model was implemented in commercially-available software, using a detailed free radical based kinetic scheme covering the components of industrial relevance. The model framework allows extension and tuning of the cracking kinetics based on available data. The parameters associated with the coking kinetic model can be estimated from plant data.

Once validated against plant data, the model serves as a valuable tool to predict KPIs such as conversion, yields and coking rates. If the model is deployed online, it can act as a soft sensor providing continuous outputs of these KPIs for each pass of the cracking furnace and support APC applications without the need of expensive gas analysers. The model can also be used for optimisation of the cracker operation to improve the VCM plant economics.

The above mentioned model has successfully been used by SCG Chemicals, Thailand to model and optimise their EDC crackers. The cracking and coking kinetic models were tuned to plant data and the validated model is then used to perform optimisation of the EDC cracker operation. The optimisation results identify potential for significant savings due to operation at higher conversion, without adversely affecting key impurity yields and run length. Optimisation results were implemented in a plant trial which helped to increase the conversion by 0.5%. Operational changes targeting further increase in conversion is envisaged after addressing the necessary plant modifications.

6. Nomenclature

R	Revenue
C	Cost
Rl	Run length
SD	Shut down time
fg	Fuel gas
dc	Decoking

References

[1] B. S. Choi, J. S. Oh, S. W. Lee, H. Kim and J. Yi, "Simulation of the Effects of CCl 4 on the Ethylene Dichloride Pyrolysis Process," *Ind. Eng. Chem. Res.,* vol. 40, pp. 4040 - 4049, 2001.

[2] A. G. Borsa, A. M. Herring, T. McKinnon, R. L. McCormick, S. Yamamoto, Y. Teraoka and Y. Natori, "Characterization of Coke Formed in Vinyl Chloride Manufacture," *Ind. Eng. Chem. Res.,* vol. 38, pp. 4259 - 4267, 1999.

[3] A. G. Borsa, A. M. Herring, T. McKinnon, R. L. McCormick and G. H. Ko, "Coke and Byproduct Formation during 1,2-Dichloroethane Pyrolysis in a Laboratory Tubular Reactor," *Ind. Eng. Chem. Res.,* vol. 40, pp. 2428 - 2436, 2001.

512815 Single & Multi-Objective Optimizations Using Parallelized Process Simulators

Abstract:

Equation-oriented (EO) optimization methods are difficult to apply to rigorous thermodynamic methods, discontinuous variables, or systems where the number of equations is not constant. Despite these challenges, EO methods are still quite commonly used for process optimization, due in part to a perception that EO methods are faster than process simulation-based methods.

By integrating parallel processing techniques into modern process simulation and optimization, rapid results can be achieved comparable to the speed of more traditional EO modeling methods. These speeds can be attained even while using first-principle models, without risking validity of the model through too many simplifying assumptions. In this presentation, several process examples will be considered to explore the advantages of using simulators for optimization projects.

513267 Integration of Computational Dynamics, Data Analysis Tools and Process Simulation for Modeling of Waste Water Treatment Plants

Abstract:

On-line and off line plant data, process flowsheet simulators and CFD are synergistically integrated for development of operating policies for waste water treatment plants. The waste water plants have long residence times, highly variable influent rates plus their compositions, and complex kinetics as well as hydrodynamics. The data gaps and flowsheet model shortcomings were consolidated using computational fluid dynamics while influent data analysis utilized neural nets. The methodology is applied to development of operating policies of a Hillsborough County waste water treatment plant.

512936 Comparative Process Analysis Techniques Applied to Four Propylene Oxide Manufacturing Routes

Abstract:

Comparative Process Analysis Techniques Applied to Four Propylene Oxide Manufacturing Routes

Dr. John C. Prindle, Jr., P.E.

Torchwood Enterprises, LLC

Several chemistries are being practiced commercially to manufacture propylene oxide. These include the conventional routes via t-butyl hydroperoxide, ethylbenzene hydroperoxide (Shell's SMPO process), hydrogen peroxide (BASF/DOW), and the proposed direct propylene epoxidation using hydrogen/oxygen mixtures. The challenges of staying competitive in this market require an understanding of the advantages/disadvantages for both your and your competitors' processes.

Comparative analysis of the commercial-scale processes using an available simulation package is one way of gaining this understanding. This task, however, contains major obstacles. Reaction kinetic models have been reported for many of the reactions practiced and proposed. However, many are heterogeneous multi-site models and cannot be directly entered into any of the simulation packages. Even though a statistical range is typically reported for each kinetic parameter in these models, how will the acceptable variation in these parameters impact downstream separations? Are there better separation alternatives which might be less sensitive to these parameter variations? This presentation will discuss the advantages and challenges of applying ASPEN Plus® to the simulation of a commercial-scale process for each of the four routes. Several findings from this comparative analysis will be highlighted.

512725 Application and Challenges of Numerical Simulations in Modeling and Scale-up of Circulating Fluidized Bed Reactors

Abstract:

Circulating fluidized bed (CFB) reactors have a long history of commercial applications ranging from coal gasification and Fluid catalytic cracking (FCC), to newer applications such as catalytic conversion of methanol to olefin (MTO) and even CO_2 Capture using solid sorbents.

Like any other technology, there are challenges in the way of scale-up, deployment and optimization of this technology, based on the lab or bench scale studies. On the other hand, despite of the long history of circulating fluidized bed operations in mature technologies such as Fluid Catalytic Cracking (FCC), complete understanding of the origin and nature of the inherently complex flow structures observed in these devices is still lacking. Adding chemical reactions to the picture makes it even more complex. Therefore, beside experimental studies, advanced and rigorous numerical modeling, i.e. computational fluid dynamics (CFD), is essential to shed light on the complex behavior and flow structure in these systems and to bridge the gap between lab/bench scale and large commercial scales.

In this talk, the strengths and shortcomings of the existing mono-disperse continuum models in simulations of fluidized bed reactors will be reviewed. The needed improvements in the inter-phase exchange properties, frictional viscosity and considering the effect of polydispersity through a novel mathematical approach, will also be discussed.

512776 Thermodynamic Modeling for CO_2 Absorption in Aqueous Potassium Carbonate Solution with Electrolyte NRTL Model

Abstract:

Potassium carbonate has recently gained attention and is considered favorable for CO_2 removal due to numerous reasons such as low economic costs and energy consumption, low volatilization and degradation and less environmental impacts compared to amine solutions. Therefore, we develop a comprehensive thermodynamic model for the $K_2CO_3 - H_2O - CO_2$ system using the Electrolyte Nonrandom Two-liquid (eNRTL) activity coefficient model in conjunction with PC-SAFT equation of state. Vapor-liquid equilibrium (VLE), heat capacity, excess enthalpy, mean ionic activity coefficient and osmotic coefficient data are used to determine eNRTL temperature dependent binary interaction parameters for both $K_2CO_3 - H_2O$ binary and $K_2CO_3 - H_2O - CO_2$ ternary systems. The model accurately represents all the thermodynamic properties and covers a wide temperature range from 273.15 to 473.15 K and CO_2 loading from 0 to 1 The model can be merged with other amine models for mixed K_2CO_3 – amine solutions and the model can be used for comparison with other CO_2 capture chemical solutions. The model should be a useful tool in the process modeling, simulation and design of CO_2 capture processes involving hot, concentrated K_2CO_3 solutions at elevated temperatures and pressures.

512813 Single and Multi-Objective Optimizations Using Parallelized Process Simulators

Abstract:

Equation-oriented (EO) optimization methods are difficult to apply to rigorous thermodynamic methods, discontinuous variables, or systems where the number of equations is not constant. Despite these challenges, EO methods are still quite commonly used for process optimization, due in part to a perception that EO methods are faster than process simulation-based methods.

By integrating parallel processing techniques into modern process simulation and optimization, rapid results can be achieved comparable to the speed of more traditional EO modeling methods. These speeds can be attained even while using first-principle models, without risking validity of the model through too many simplifying assumptions. In this presentation, several process examples will be considered to explore the advantages of using simulators for optimization projects.

512830 Thermodynamic Modeling of the $LiCl+LiBr+LiNO_3+LiI+H_2O$ System with the Electrolyte NRTL Model

Abstract:

Thermodynamic models are essential to facilitate the advancement of process design, optimization, and operation range of absorption heat pumps and refrigeration units that utilize lithium salt systems. The lithium bromide-water solution is a popular working fluid due to its environmentally friendly, nonvolatile, and non-toxic properties; however, this solution has issues with corrosion and crystallization. One proposed solution is to modify the lithium bromide-water working solution by adding anticrystallization and anticorrosion agents such as lithium chloride, lithium iodide, and lithium nitrate. The objective of the present work is to develop a comprehensive thermodynamic representation of the $LiCl+LiBr+LiNO_3+LiI+H_2O$ quinary solution by implementing the symmetric electrolyte NRTL (eNRTL) activity coefficient model. This solubility model correlates the nonideality of the solution with the composition dependency by utilizing two binary interaction parameters for each of the electrolyte-electrolyte and molecule-electrolyte pairs in the system. Each parameter is determined by applying up to three temperature dependent coefficients and then regressed from experimental data. This allows the parameters to establish the accuracy of the model for concentrations up to saturation and temperatures between 273.15-473.15 K.

513562 Efficient Numerical Optimization of an Aerated Mixing Vessel for Fermentation

Abstract:

Oxygen is a critical component of aerobic fermentations and is sparingly soluble in aqueous solutions. Due to the slow mass transfer of oxygen from the bubbles into the liquid the rate of cell metabolism depends on the rate of oxygen supply from the gas phase. Mass transfer across gas-liquid interfaces plays also a vital role in many other reaction systems. The mass transfer is a function of the gas-liquid surface area which is a function of the coalescence and breakup rate which is again a function of the local velocity and turbulence. The local flow condition are finally determined by the reactor geometry, the fluid properties and â€“ in this study â€“ by the rotational speed of the impeller. So from a reactor performance standpoint, high aeration rates and high agitation rates are favorable.

On the other hand the operating costs are mainly determined by the total power consumed for the aeration and by the impeller of an aerated mixing vessel and should be as low as possible.

Traditionally dimensionless correlations and expensive experimental investigations are used to design stirred tank fermenters in production scale. However, design methodologies available for stirred tank fermenters are often insufficient and are using simplifications which affects e.g. the mass transfer significantly. Using computational fluid dynamics (CFD) based on validated models enables using fewer assumptions on the geometry and physics resulting in a better insight into the system.

In this work the performance of a stirred tank fermenter is going to be optimized using CFD simulations in combination with design explorations. After validating some implemented models especially to incorporate bubble size distribution including coalescence and breakup in the finite volume code STAR-CCM+, HEEDS is used to explore the design space. HEEDS is a multi-objective optimization tool using hybrid, adaptive search strategies to find better designs, faster and couples seamlessly to STAR-CCM+, so that no model simplifications are needed. Details how this approach works and how it can be applied to other problems will be shown. Finally the obtained Pareto front w.r.t. power consumption and mass transfer rate will be presented.

513225 Multi-Oil Product Pipeline Scheduling with Appropriate Handling of Oil Mixing Interface

Abstract:

Pipeline is more efficient and reliable to transfer large amounts of petroleum products than other ways via road, railroad, and vessel. To save capital and operating costs, one pipeline needs to transfer different types of oil products to different oil depots according to their ordinary demands. As two different consecutive oil products inside a pipeline will not be separated by a physical divider, the product mixing is inevitable near the interface of two oil products. To guarantee the oil quality adopted by storage tanks at depots, a considerable amount of mixing oil product from the pipeline needs to be dealt with properly at depots. In practice, three kinds of methods are used. The first one is to blend mixture oils into pure product, which might have some oil quality problems; the second one is to separate the oil mixture into pure oil products through distillation; and the third one is the hybrid method coupling the former two methods together. Conceivably, the multiproduct pipeline scheduling problem with consideration of dealing with oil mixing inside the pipeline present a very challenging problem to industries. There are few works reported in this area (e.g., Rejowski et al. (2003); Cafaro and Cerdá (2004 and 2010)). However, they still lack studies on dealing with the oil mixture product.

In this work, a continuous-time based MILP model is developed to optimize the transportation and distribution of multi oil products, which includes a long-distance pipeline, several depots with storage tanks and mixing tanks dedicated to hold pure product and mixture respectively. The oil mixing interface inside the pipeline is modeled as three parts: two product-rich mixing parts on both sides of the mixing interface and one well mixing part in the middle. The hybrid method for blending product-rich mixture into the pure product and refining the well mixing mixture has been adopted. Overall, the scheduling model could provide optimal solutions including product feeding into pipeline, product movements inside pipeline, discharging product to multiple depot tanks, as well as sending out oil product based on consumer demands in the continuous time formulation base. The efficacy of the developed scheduling model is demonstrated by multiple case studies.

Keywords: Multiproduct Scheduling, Pipeline, Mixing Product Management, MILP

References:

Cafaro, D., & Cerdá, J. (2004). Optimal scheduling of multiproduct pipeline systems using a non-discrete MILP formulation. Computers and Chemical Engineer- ing, 28, 2053–2068. ISSN 0098-1354.

Cafaro, D., & Cerdá, J. (2010). Operational scheduling of refined products pipeline networks with simultaneous batch injections. Computers and Chemical Engineer- ing, 34, 1687–1704 .

Rejowski, R., & Pinto, J. (2003). Scheduling of a multiproduct pipeline system. Com- puters and Chemical Engineering, 27, 1229–1268 .

513699 Advanced Model-Based Dynamics Optimization of a Commercial Natural Gas Recovery Unit

Abstract:

Natural gas liquids (NGLs) are the main contributor to the profitability of natural gas processing plants. During the recent period of relatively low gas prices NGLs spot prices have experienced a slower drop remaining relatively steady. Therefore, driving operating plants towards higher NGL production can lead to a more profitable position. In this work, we look to use model-based optimization on a selection of NGL plants to optimize them towards higher profitability and/or lowest energy consumption.

Such studies are meaningful only when based on sound, high-fidelity models of the plant, ensuring that â€œoptimalâ€ solutions obtained satisfy all important safety and operability constraints. In the typical workflow a model is built in commonly used flowsheeting tools based on the Sequential Modular approach. Limitations in this approach result in engineers having to resort to manual trial-and-error simulations with no guarantee of obtaining an optimal solution or even one that satisfies all the important constraints. In this work models are being built in an Equation Oriented environment which allows the use of formal mathematical optimization methods to guarantee an optimal solution satisfying all constraints is found.

In this paper we use the example of an on-going study of a NGL Plant in Abu Dhabi region to explain the practical benefits of Advanced Model-based Optimization. We will describe and show how the key first step in such a study is validation: ensuring that model predictions of key results are matching the actual plant behavior. We will describe how the model is then used to understand the best operational strategies for the facility through model-based optimization. This demonstrates how the approach used overcomes the limitations of the typical workflow and models used. In particular, the use of tools based on the Sequential Modular approach is limited in optimization capabilities due to its problematic handling of non-standard specifications, closing of recycle loops and the unavailability of gradients. The trial-and-error simulations used in this approach are not only time consuming and error prone but often limited in the operating decisions and constraints that can be considered. We will show how using an Equation Oriented environment overcomes these limitations and allows an optimal solution satisfying all operational and safety constraints to be found using mathematical optimization methods.

Finally we will show how this leads to the set-up of operating guidelines to allow operators to drive the plant towards optimal operation; giving them the ability to respond to changes in the plant or economic conditions quicker and with more confidence than previously possible. In addition to moving the plant towards optimal operation, the models will also provide the ability to address a range of operational situations.

513652 CFD Simulation of Gas-Liquid Flow in Monolith Channels

Abstract:

Multiphase monolith reactors are a kind of novel process intensification reactors which are promising for many catalytic gas-liquid-solid reactions encountered in chemical, petrochemical, biochemical, and environmental processes. Multiphase hydrodynamics and distribution characteristics in a monolith bed are the important factors which affect gas-liquid mass transfer in monolith channels significantly and therefore the whole reactor performance. In this paper, a two-dimensional (2D) CFD model has been built for analyzing the distribution characteristics of gas-liquid flow within a monolith bed with 0.048m diameter and 400 cpsi. The volume of fluid (VOF) multiphase flow model is implemented to calculate the flow field and capture the interface between the gas and liquid phases then obtain hydrodynamic parameters in each channels of the monolith bed. The simulation results matched well with the experimental results presented by Zhou (Yuan Zhou, et al, 2014, AICHE Journal).

AUTHOR INDEX